//
Pluto

The Exiled Wanderer

JD ARDEN

Preface: A Planet, Then Not

Pluto, once the ninth planet of the solar system and celebrated as the outermost member of our celestial family, has endured a tumultuous journey in the public imagination. Discovered in 1930 by Clyde Tombaugh, it was hailed as a triumph of astronomical prediction and a testament to humanity's ability to uncover the unseen. For decades, Pluto held its place in school textbooks and inspired awe as the lonely, icy outpost of the solar system. Yet, in 2006, a controversial decision by the International Astronomical Union (IAU) reclassified Pluto as a dwarf planet, stripping it of its planetary status and igniting debates that continue to this day.

This demotion did not diminish Pluto's allure; if anything, it deepened its mystique. Pluto became the exiled wanderer, a world that defied easy categorization and reminded us of the fluidity of scientific understanding. Its unique position at the edge of the solar system, its complex geology, and its symbolic ties to both mythology and human exploration cemented its place in the public's imagination.

The arrival of NASA's New Horizons spacecraft in 2015 was a watershed moment. For the first time, humanity glimpsed Pluto's surface in breathtaking detail. Far from being a frozen, inert world, Pluto revealed itself to be a place of astonishing activity and complexity. Its icy plains, towering mountains, and potential cryovolcanism shattered expectations and redefined what we thought we knew about distant worlds.

Pluto's story is one of change—geological, scientific, and cultural. It challenges us to think beyond rigid definitions and embrace the diversity of the solar system. As the exiled wanderer, it symbolizes the tension between belonging and exclusion, the yearning for discovery, and the enduring allure of the unknown.

This book explores Pluto in all its facets: its scientific significance, its cultural impact, and its philosophical implications. From the icy heart of Tombaugh Regio to the debates surrounding its planetary status, Pluto offers a lens through which to examine not only the outer reaches of the solar system but also humanity's evolving relationship with exploration, identity, and discovery.

Pluto

Chapter 1: Pluto's Icy Heart

When the New Horizons spacecraft sent back the first close-up images of Pluto in 2015, one feature stood out immediately: a vast, heart-shaped plain of nitrogen ice, gleaming against the dark, rugged terrain that surrounded it. This region, named Tombaugh Regio after Pluto's discoverer, Clyde Tombaugh, became an emblem of Pluto's unexpected complexity and vitality. Far from being a static, frozen wasteland, Tombaugh Regio revealed Pluto as a world with active geology and dynamic processes, defying expectations for a small, distant body in the Kuiper Belt.

Tombaugh Regio consists of two distinct lobes. The western lobe, called Sputnik Planitia, is a vast, smooth expanse of nitrogen, methane, and carbon monoxide ice. This plain is remarkable for its lack of impact craters, indicating that its surface is geologically young—likely no more than 10 million years old. Such youthfulness is a striking discovery for a world that resides in the frigid outer reaches of the solar system, where geological activity was thought to have ceased billions of years ago.

The surface of Sputnik Planitia is marked by a network of polygonal cells, ranging from 16 to 40 kilometers across. These cells are the result of convection, a process driven by heat escaping from Pluto's interior. The nitrogen ice, warmed slightly from below, rises in the center of these cells and cools as it sinks along their edges, creating a dynamic, churning landscape. This process, akin to the bubbling of a pot of boiling water, reshapes the surface over time, erasing craters and maintaining its youthful appearance.

Surrounding Sputnik Planitia are towering water-ice mountains, some rising as high as 6 kilometers (nearly 4 miles). These mountains, composed of frozen water as hard as rock at Pluto's temperatures, serve as barriers confining the nitrogen ice of the plain. Their rugged, angular shapes suggest a history of tectonic activity, with forces pushing and pulling at Pluto's crust over eons.

The eastern lobe of Tombaugh Regio is more varied, featuring rougher terrain interspersed with smoother areas. This region hints at the interplay of multiple processes, including glaciation, sublimation (where ice transitions directly to vapor), and deposition of volatile compounds. The boundary between the two lobes is sharply defined, reflecting the complex interactions between Pluto's various geological systems.

The active geology of Tombaugh Regio raises fundamental questions about the energy sources driving these processes. Pluto, with its small size and great distance from the Sun, was expected to be a geologically dead world, its internal heat long dissipated. Yet the evidence suggests otherwise. One possibility is that Pluto retains heat from its formation, supplemented by the decay of radioactive elements within its rocky core. Another hypothesis is that interactions with Charon, Pluto's largest moon, generate tidal forces that contribute to internal heating, though this effect is likely minimal given the current state of their orbits.

The composition of Sputnik Planitia also highlights Pluto's unique chemistry. The nitrogen, methane, and carbon monoxide ices that dominate the plain are volatile at Pluto's surface temperatures, meaning they can sublimate into the thin atmosphere and recondense elsewhere. This cycle of sublimation and deposition creates a dynamic environment where the landscape is continually reshaped, despite the planet's frigid conditions.

In addition to its scientific significance, Tombaugh Regio carries deep symbolic weight. Its heart shape, visible even in low-resolution images, immediately captured the public's imagination, transforming Pluto from an abstract celestial object into a world with personality and charm. The region's naming after Clyde Tombaugh further cements its place in the narrative of discovery, honoring the astronomer who first brought Pluto to light.

The discovery of Tombaugh Regio and its active geology underscores the importance of exploration. Before New Horizons, Pluto was little more than a distant speck, its surface features indistinguishable and its nature uncertain. The spacecraft's flyby revealed not just a world, but a story—a dynamic, evolving planet that continues to defy expectations and challenge assumptions about what is possible in the outer solar system.

Pluto's icy heart is a reminder of the power of curiosity and the surprises that await us when we venture into the unknown. It is a testament to the complexity and diversity of the solar system, showing that even the

Pluto

smallest and most distant worlds can be full of wonder. In Tombaugh Regio, we see a reflection of our own drive to explore, to uncover the hidden, and to find meaning in the vastness of space.

Chapter 2: A Dwarf Among Giants

Pluto occupies a unique position in the solar system, not only geographically but also conceptually. Once hailed as the ninth planet, it is now classified as a **dwarf planet**, a term that places it alongside a growing family of celestial bodies in the Kuiper Belt. This reclassification, while controversial, highlights Pluto's role as a bridge between the traditional planets and the icy frontier beyond. It is a world caught between definitions, an intermediary in both scale and significance—a dwarf among giants, yet a giant among its peers.

The Kuiper Belt, a vast region of icy bodies beyond Neptune, is Pluto's true home. Discovered in the late 20th century, the Kuiper Belt has transformed our understanding of the outer solar system. It is a realm of thousands, perhaps millions, of objects ranging from small, icy fragments to dwarf planets like Pluto, Eris, and Haumea. These objects are remnants from the early solar system, left largely undisturbed in the cold, dark reaches beyond the planets.

Pluto stands out within the Kuiper Belt not only because of its size but also because of its complexity. With a diameter of about 2,377 kilometers (1,477 miles), it is the largest known Kuiper Belt object (KBO) and among the most geologically active. While other KBOs, like Eris and Haumea, boast intriguing features such as elongated shapes or surface variations, none rival Pluto's combination of geological diversity, atmospheric activity, and moon system. This makes Pluto a keystone for understanding the Kuiper Belt as a whole.

Pluto's orbit further distinguishes it within the Kuiper Belt. Unlike the nearly circular orbits of the traditional planets, Pluto's path around the Sun is elliptical and highly inclined, tilted about 17 degrees relative to the plane of the solar system. This eccentric orbit occasionally brings Pluto closer to the Sun than Neptune, a reminder of the chaotic interactions that shaped the solar system's architecture. Despite this apparent overlap, Pluto and Neptune never collide due to their **orbital resonance**: for every three orbits Neptune completes, Pluto completes two, a relationship that keeps the two bodies safely apart.

This resonance highlights Pluto's dynamic relationship with the giants of the solar system, particularly Neptune. While Pluto's size and mass are dwarfed by Neptune's, the gravitational interplay between the two has profound implications. It is likely that Neptune's migration during the early solar system helped scatter objects into the Kuiper Belt, including Pluto. This migration may also explain Pluto's orbital characteristics, as the gravitational interactions with Neptune could have captured Pluto into its current resonance.

Despite its diminutive size compared to the eight planets, Pluto wields significant influence within its own system. It boasts five known moons, the largest of which, **Charon**, is so massive relative to Pluto that the two are often described as a binary system. The gravitational relationship between Pluto and Charon creates a unique dynamic, with both bodies orbiting a shared center of mass that lies outside Pluto itself. This binary nature sets Pluto apart from other Kuiper Belt objects, further emphasizing its role as a standout among its peers.

The discovery of other dwarf planets in the Kuiper Belt, such as Eris, Haumea, and Makemake, has reshaped our understanding of Pluto's place in the solar system. These objects share many characteristics with Pluto, including size, composition, and orbital traits, suggesting that Pluto is not an outlier but part of a broader population. This realization led to the controversial decision in 2006 by the International Astronomical Union (IAU) to redefine what constitutes a planet, resulting in Pluto's reclassification as a dwarf planet.

While this decision sparked widespread debate and public outcry, it also underscored the evolving nature of science. The demotion of Pluto was not a diminishment of its importance but an acknowledgment of the growing diversity of objects in the Kuiper Belt. It recognized that Pluto is not unique in its size or location but remains unique in its complexity and historical significance.

Philosophically, Pluto's position as a "dwarf among giants" invites reflection on the nature of categorization and the human tendency to create boundaries. The debate over Pluto's status highlights the tension between simplicity and complexity, between defining a category and embracing the diversity within it. Pluto challenges us to think beyond rigid definitions and to appreciate the spectrum of celestial bodies that populate the solar system.

Pluto's status as both a dwarf planet and a Kuiper Belt object (KBO) also raises questions about what it means to belong. It exists on the fringe of two worlds: the planetary realm, with its eight giants, and the Kuiper Belt, with its multitude of icy bodies. In this liminal position, Pluto becomes a symbol of transition and connection, a reminder that boundaries in the universe are often fluid and overlapping.

The exploration of Pluto by NASA's **New Horizons mission** has further solidified its significance within the Kuiper Belt. The detailed images and data sent back by the spacecraft revealed a world far more dynamic than expected, with flowing glaciers, towering mountains, and a tenuous atmosphere that interacts with the solar wind. These discoveries not only expanded our understanding of Pluto but also provided a framework for studying other KBOs. By examining Pluto, we gain insights into the processes that shaped the Kuiper Belt and the early solar system.

Pluto's position as a dwarf among giants is both a challenge and an opportunity. It reminds us of the solar system's vastness and diversity, urging us to look beyond traditional categories and explore the full range of celestial phenomena. At the same time, Pluto's complexity and beauty reaffirm that even the smallest and most distant worlds can hold profound significance.

As we continue to study the Kuiper Belt and its inhabitants, Pluto remains a touchstone, a guide to understanding a region of the solar system that is only beginning to reveal its secrets. It is a world that defies expectations and invites exploration, standing as both an outlier and a keystone in the story of the solar system.

Pluto's journey—from its discovery to its reclassification, and from its icy heart to its role as a KBO—reminds us that the universe is not easily categorized. It is a place of wonder and complexity, where even the smallest objects can inspire the grandest questions.

Chapter 3: Pluto's Companion: Charon

Among Pluto's five known moons, **Charon** stands apart as an exceptional companion. Discovered in 1978 by astronomer James Christy, Charon is not just another satellite; it is a celestial partner that shapes and is shaped by Pluto in profound ways. With a diameter of about 1,212 kilometers (753 miles), Charon is roughly half the size of Pluto and so massive relative to its host that the two bodies orbit a shared center of gravity, or barycenter, that lies outside Pluto itself. This relationship makes Pluto and Charon unique in the solar system, often referred to as a **binary dwarf planet system**.

Charon's size and proximity to Pluto create a gravitational bond that dominates their interaction. Unlike most planetary systems, where moons orbit around a central body, Pluto and Charon revolve around a barycenter that hovers in space between them. This dynamic gives the appearance of a cosmic dance, with each body pulling on the other in a delicate equilibrium. The result is a system where both Pluto and Charon are tidally locked to each other, always presenting the same face toward one another. For an observer on Pluto, Charon would remain fixed in the sky, an ever-present companion.

The surface of Charon is as intriguing as its relationship with Pluto. Images captured by the **New Horizons spacecraft** in 2015 revealed a world of surprising geological diversity and activity. Charon's northern hemisphere is dominated by a dark reddish region informally called **Mordor Macula**, likely composed of tholins—organic compounds formed by the interaction of sunlight and methane or nitrogen gases. These compounds may originate from Pluto's atmosphere, which escapes into space and settles on Charon's surface, creating a striking contrast with the lighter, icy terrain below.

Elsewhere, Charon's surface is marked by vast plains, canyons, and fractures. The most prominent feature is a system of deep chasms that stretch across its equator, dwarfing even the Grand Canyon on Earth. These chasms, some of which are several kilometers deep, suggest a history of internal activity, possibly driven by the cooling and contraction of Charon's subsurface layers.

One of the most compelling discoveries about Charon is the evidence of **cryovolcanism**—a process where icy materials, rather than molten rock, erupt from the interior. The smooth plains on Charon's surface, such as the **Vulcan Planitia**, may have been formed by eruptions of ammonia-rich water that flowed and froze, creating a resurfaced landscape. This evidence of geological activity indicates that Charon's interior once retained enough heat to drive such processes, challenging the assumption that small, icy worlds are geologically inert.

The origins of Charon remain a topic of scientific investigation. The prevailing hypothesis suggests that Charon was formed as a result of a massive collision early in Pluto's history. In this scenario, a proto-Pluto was struck by a large object, ejecting debris that coalesced to form Charon and possibly Pluto's smaller moons—Styx, Nix, Kerberos, and Hydra. This hypothesis is supported by the similarities in composition between Pluto and Charon, as well as their tight gravitational relationship.

Charon's existence has profound implications for understanding Pluto. The tidal interactions between the two bodies have played a significant role in shaping their orbits, rotation rates, and even internal structures. For example, tidal forces likely contributed to the redistribution of mass within Pluto, influencing its surface features and internal dynamics. Similarly, Charon's surface bears the imprint of its gravitational relationship with Pluto, including the possibility of tidal heating during its early history.

The binary nature of the Pluto-Charon system also provides a unique laboratory for studying the dynamics of celestial bodies. The shared barycenter and the tidal locking between Pluto and Charon are rare phenomena that offer insights into how gravitational interactions shape the evolution of planetary systems. These dynamics are not confined to the Pluto-Charon system; they have analogs in exoplanet systems and other binary objects in the Kuiper Belt, making this pairing a valuable reference point for broader astronomical studies.

Pluto

Philosophically, the relationship between Pluto and Charon invites reflection on the nature of companionship and interdependence. Unlike the hierarchical relationships seen in most planetary systems, where a dominant planet governs its satellites, Pluto and Charon exist in a more egalitarian partnership. Each body influences the other, their fates intertwined in a mutual orbit that reflects the balance of forces at play. This dynamic serves as a metaphor for interconnectedness, a reminder that even in the vastness of space, relationships define and shape existence.

Charon's role as Pluto's companion extends beyond their physical interaction. It enhances Pluto's significance in the solar system, adding complexity and depth to its story. The binary nature of the Pluto-Charon system challenges traditional definitions of planets and moons, pushing the boundaries of how we categorize celestial objects. It underscores the diversity of systems that exist in the cosmos, where even the smallest and most distant worlds can defy expectations and reveal new paradigms.

The exploration of Charon by New Horizons marked a turning point in our understanding of this moon. Before the mission, Charon was little more than a faint dot in telescopes, its nature and surface features unknown. The detailed images and data returned by New Horizons transformed it into a world of its own, one with a rich geological history and an enduring connection to its host.

Charon is more than a moon; it is a partner, a co-architect of the Pluto system, and a key to understanding the dynamics of the Kuiper Belt. Its existence challenges us to think about celestial relationships in new ways, emphasizing the importance of connections and the interplay of forces that shape worlds.

In Charon, we find a reminder of the unexpected—of how even the most distant and seemingly simple objects can hold complexity, beauty, and meaning. It is a testament to the richness of the solar system and a call to continue exploring, questioning, and imagining what lies beyond.

Chapter 4: Cryovolcanoes and Icy Landscapes

Pluto's surface is a masterpiece of icy geology, a canvas painted with towering mountains, sprawling plains, and rugged canyons. Among its most surprising and significant features are the evidence of **cryovolcanism** and the variety of icy landscapes that defy expectations for a world so small and distant from the Sun. These discoveries, made during the **New Horizons** flyby in 2015, have transformed our understanding of Pluto, revealing it as an active and dynamic world with a history shaped by internal processes as much as external forces.

Cryovolcanism, or **cold volcanism**, occurs when icy materials erupt from a planet or moon's interior rather than molten rock. On Pluto, cryovolcanoes eject water, ammonia, nitrogen, and methane in a semi-liquid or vaporized state, which then freezes upon reaching the surface. This phenomenon is remarkable because it indicates that Pluto retains internal heat—despite its small size and distant location—that drives these processes.

One of the most striking examples of cryovolcanism on Pluto is the feature informally named **Wright Mons**, a massive dome-like structure rising 4 kilometers (about 2.5 miles) above the surrounding terrain. Wright Mons is one of the largest cryovolcanic structures in the solar system, comparable in size to Hawaii's Mauna Loa, though composed of vastly different materials. Its surface is pocked with depressions, likely the remnants of past eruptions, and covered with a mixture of water ice and exotic volatiles such as ammonia, which lowers the freezing point of water and facilitates the flow of cryolava.

Nearby lies **Piccard Mons**, another potential cryovolcano with similar characteristics. Together, Wright Mons and Piccard Mons suggest a network of cryovolcanic activity, indicating that Pluto's interior has been geologically active in the relatively recent past. The smooth and sparsely cratered terrain surrounding these features further supports the idea that cryovolcanic eruptions have reshaped the surface, erasing older impacts and creating new landscapes.

Pluto

The presence of cryovolcanoes raises fundamental questions about the heat sources that drive these processes. Conventional wisdom suggests that small, icy bodies like Pluto should have long since cooled and become geologically inert. Yet the evidence of cryovolcanism indicates that Pluto retains enough internal heat to sustain subsurface reservoirs of liquid or semi-liquid materials. This heat may originate from the **decay of radioactive isotopes** in Pluto's rocky core, a process that generates energy over billions of years. Alternatively, Pluto's early history—marked by intense impacts and tidal interactions—may have left residual heat that persists today.

Pluto's icy landscapes are equally compelling, showcasing a variety of terrains that reflect the interplay of geological, atmospheric, and cryovolcanic processes. The most iconic of these is **Sputnik Planitia**, the western lobe of Tombaugh Regio, a vast plain of nitrogen ice that exhibits polygonal cells created by convection. The smoothness of Sputnik Planitia's surface, devoid of significant cratering, indicates ongoing resurfacing driven by the slow, churning motion of nitrogen ice.

Beyond Sputnik Planitia lies a world of rugged mountains and fractured terrains. The **water-ice mountains**, such as the Norgay Montes and Hillary Montes, rise to heights of 3 to 6 kilometers (about 2 to 4 miles), rivaling some of the tallest peaks on Earth. These mountains are thought to be composed of water ice, which is as hard as rock at Pluto's frigid temperatures. Their angular shapes suggest tectonic forces at work, possibly driven by the expansion and contraction of Pluto's icy crust as it responds to internal and external forces.

Another fascinating feature of Pluto's icy landscapes is the presence of **bladed terrain**, found in regions like Tartarus Dorsa. These sharp ridges, rising hundreds of meters above the surface, are thought to form through sublimation, where methane ice transitions directly from solid to vapor under Pluto's thin atmosphere. This process leaves behind jagged formations that resemble enormous blades or fins, creating a surreal and alien landscape.

Pluto's cryovolcanic and icy features are not isolated phenomena; they interact with the planet's tenuous atmosphere and its subsurface layers, creating a dynamic system that evolves over time. The atmosphere, composed primarily of nitrogen with traces of methane and carbon monoxide, plays a role in redistributing volatiles across the surface. Sublimation and deposition cycles cause materials to migrate, shaping the terrain and creating features such as frost-covered plains and seasonal deposits of methane ice.

The possibility of a **subsurface ocean** beneath Pluto's icy crust adds another layer of intrigue. If such an ocean exists, it could be a source of the cryovolcanic activity observed on the surface, providing the liquid materials needed for eruptions. The presence of ammonia, detected in the cryolava of Wright Mons, supports this idea, as ammonia acts as an antifreeze that helps maintain liquid water under extreme conditions. A subsurface ocean on Pluto would not only explain its geological activity but also raise the tantalizing possibility of habitability, albeit in forms very different from those on Earth.

Philosophically, Pluto's cryovolcanoes and icy landscapes challenge our assumptions about what small, distant worlds can be. Before New Horizons, Pluto was often imagined as a static, frozen relic of the Kuiper Belt, a world locked in cosmic stasis. The reality revealed by the spacecraft is far more dynamic: a planet of ongoing change, where ice flows like lava, mountains rise and erode, and the surface is continually reshaped by forces we are only beginning to understand.

Pluto's icy landscapes also remind us of the diversity of geological processes across the solar system. While Earth's volcanoes are fueled by molten rock and tectonic activity, Pluto's cryovolcanoes and nitrogen-ice plains demonstrate that even the coldest and most remote environments can harbor geological activity. This diversity reflects the adaptability of nature, which finds ways to create and transform in even the most unlikely conditions.

As we look to the future, the study of Pluto's cryovolcanic and icy features offers a roadmap for exploring other icy worlds in the Kuiper Belt and beyond. By understanding the processes that shape Pluto, we gain insights into the broader dynamics of the outer solar system and the conditions that may exist on similar bodies in distant planetary systems.

Pluto's cryovolcanoes and icy landscapes are not just features of a distant world; they are testaments to the enduring creativity of the cosmos. They remind us that even at the edges of the solar system, where sunlight is

Pluto

faint and temperatures plummet, there is activity, beauty, and the potential for discovery.

Chapter 5: The Legacy of New Horizons

The **New Horizons spacecraft**, launched by NASA in January 2006, revolutionized our understanding of Pluto and its place in the solar system. After a nine-and-a-half-year journey spanning more than 5 billion kilometers (3 billion miles), New Horizons became the first mission to visit Pluto, offering humanity its first close-up view of this distant world. The data and images it returned transformed Pluto from a faint speck in telescopes into a dynamic and complex planet, reshaping scientific paradigms and capturing the public's imagination.

Before New Horizons, Pluto was largely a mystery. Observations from Earth-based telescopes and the Hubble Space Telescope hinted at surface variations and an atmosphere, but the details were elusive. Pluto's small size, distant location, and tenuous atmosphere made it difficult to study. Many of its

most intriguing features—its geological activity, icy landscapes, and atmospheric dynamics—remained entirely hidden. The arrival of New Horizons on July 14, 2015, changed all that, offering an unprecedented glimpse of a world that exceeded all expectations.

The spacecraft's approach to Pluto began months before the closest flyby. As New Horizons closed in, its cameras and instruments captured progressively clearer images, revealing features that hinted at Pluto's complexity. These early images, showing stark contrasts between bright and dark regions, foreshadowed the surprises to come. By the time New Horizons reached its closest approach—passing just 12,500 kilometers (7,800 miles) above Pluto's surface—it was clear that this small, icy world was anything but static.

One of the most iconic discoveries of New Horizons was **Tombaugh Regio**, the heart-shaped region on Pluto's surface named after its discoverer, Clyde Tombaugh. Within this region lies **Sputnik Planitia**, a vast, smooth plain of nitrogen ice. The surface of Sputnik Planitia, marked by polygonal cells created by convection, revealed ongoing geological activity. This was a shocking discovery: a world as distant and small as Pluto was not expected to retain the internal heat necessary for such processes.

New Horizons also revealed **towering mountains of water ice**, such as those in the Norgay Montes and Hillary Montes regions. These mountains, some rising as high as 6 kilometers (4 miles), underscored Pluto's geological diversity. Their rugged shapes and icy composition hinted at a dynamic past involving tectonic forces, cryovolcanism, or the movement of subsurface materials.

The mission provided unprecedented insights into Pluto's **atmosphere**, a tenuous envelope of nitrogen, methane, and carbon monoxide that extends hundreds of kilometers above the surface. New Horizons' instruments captured stunning images of Pluto's atmospheric haze, which scatters sunlight to create a bluish glow. The detection of layers of haze and evidence of atmospheric escape to space revealed an intricate interaction between the atmosphere and the surface, driven by seasonal cycles and sublimation of volatile ices.

The spacecraft also turned its attention to Pluto's **moons**, particularly **Charon**, the largest and most intriguing. Charon, with its deep canyons, smooth plains, and dark polar cap, proved to be almost as fascinating as Pluto itself. The New Horizons data suggested that Charon, like Pluto, experienced geological activity in its past, including cryovolcanism that resurfaced parts of its terrain. The relationship between Pluto and Charon, particularly their shared barycenter and tidal locking, emerged as one of the most unique dynamics in the solar system.

Pluto's smaller moons—Styx, Nix, Kerberos, and Hydra—also received attention. New Horizons revealed them as irregularly shaped, rapidly spinning objects with surprisingly bright surfaces. These moons are thought to have formed during the same giant impact event that likely created Charon, adding another layer to Pluto's complex history.

New Horizons' discoveries extended beyond Pluto itself. The spacecraft's observations provided critical insights into the **Kuiper Belt**, the vast region of icy bodies that lies beyond Neptune. By studying Pluto's surface composition, orbit, and interactions with its moons, New Horizons helped refine models of how Kuiper Belt objects (KBOs) form and evolve. The mission demonstrated that Pluto is not an isolated anomaly but a representative of a broader population of objects, many of which remain unexplored.

Perhaps the most profound impact of New Horizons was the way it redefined **expectations for planetary exploration**. Before the mission, Pluto was often thought of as a static, frozen world—a relic of the early solar system preserved in the icy reaches of the Kuiper Belt. The discoveries of active geology, atmospheric dynamics, and a history of cryovolcanism shattered that notion. New Horizons showed that even the smallest and most distant worlds could be dynamic and full of surprises, challenging assumptions about what is possible in the outer solar system.

The mission also reignited public interest in Pluto, particularly in light of the controversy surrounding its reclassification as a dwarf planet. The images of Pluto's icy heart and rugged landscapes captured the imagination of millions, transforming the way people viewed this enigmatic world. Pluto, once the "underdog" of the solar system, became a symbol of resilience and discovery.

Scientifically, the legacy of New Horizons is far-reaching. The data it collected continues to fuel research, with scientists analyzing its findings to refine models of planetary geology, atmospheric science, and solar system dynamics. The mission also set the stage for future exploration of the Kuiper Belt, demonstrating the value of studying this distant region and its inhabitants.

Philosophically, New Horizons speaks to the power of human curiosity and the drive to explore. The spacecraft's journey to Pluto—over nearly a decade, across billions of kilometers—was a testament to perseverance and ingenuity. It reminds us that even the most remote and seemingly unreachable places in the universe can be within our grasp, given enough determination and creativity.

The mission also highlights the importance of **collaboration and vision**. The success of New Horizons was the result of decades of work by scientists, engineers, and mission planners, all united by a shared goal of uncovering Pluto's secrets. It is a reminder that exploration is not the work of individuals but of communities, working together to expand the boundaries of knowledge.

Pluto

New Horizons did more than just visit Pluto; it opened a new chapter in our understanding of the solar system. By revealing Pluto as a dynamic and complex world, the mission expanded our horizons—both literally and figuratively—and challenged us to think differently about the nature of planetary systems.

As New Horizons continues its journey into the Kuiper Belt and beyond, its legacy endures. It serves as a beacon of exploration, a symbol of humanity's capacity to reach for the stars and uncover the wonders of the universe. Pluto, once the most distant member of the solar system's planetary family, now stands as a testament to the power of discovery, its icy heart beating with the promise of new horizons yet to come.

Chapter 6: Pluto in Myth and Science

Pluto is a celestial object where myth and science converge, a symbol and a reality that spans the realms of imagination and empirical discovery. Its name, suggested by an 11-year-old girl, Venetia Burney, after the Roman god of the underworld, resonates deeply with the character of this distant world. Pluto, the god, ruled over hidden realms, wealth buried in the earth, and the transition between the living and the dead. These associations mirror the scientific Pluto: a planet at the edge of the solar system, enigmatic, elusive, and a repository of ancient knowledge about the universe's formation.

When Clyde Tombaugh discovered Pluto in 1930, the world welcomed it as the ninth planet, the farthest member of the solar system's known family. Its distance and faint light only added to its mystique. Named for a god of darkness, Pluto quickly became associated with the mysterious outer reaches of the solar system—a realm both literal and metaphorical. Yet, as much as its mythological roots celebrated its otherworldly nature, Pluto's scientific story began with practical questions, particularly the search for a "Planet X" to explain irregularities in the orbits of Uranus and Neptune. This search eventually led to the Lowell Observatory and Tombaugh's methodical work, culminating in the discovery of a world that changed how humanity understood the cosmos.

Pluto's association with the underworld god captures its dual identity as both a boundary and a bridge. In Roman mythology, Pluto presided over finality, yet his domain was also a space of transition—a place where the living and the dead crossed paths, where permanence and impermanence intertwined. Similarly, Pluto exists at a crossroads in the solar system. It is the last of the planets as traditionally imagined and the first of a vast and unexplored frontier, the Kuiper Belt. This boundary, like the underworld itself, is not a rigid wall but a porous threshold that blurs distinctions and invites exploration.

The god Pluto was also a figure of hidden riches. Ancient myths tied him to the wealth of the earth, the treasures of gold and silver buried deep beneath the surface. In a scientific sense, Pluto fulfills this role as well, serving as a repository of primordial material from the solar system's earliest days. As a Kuiper Belt object, Pluto holds the icy and rocky remnants of the nebula that gave birth to the planets. Its surface and composition offer scientists a glimpse into conditions that existed billions of years ago, making Pluto a key to understanding the origins of the solar system.

The journey of Pluto's scientific identity has been as transformative as its mythological symbolism. For decades, it held its place as the ninth planet, the mysterious outlier on the fringes of the Sun's influence. Yet discoveries in the late 20th century began to reshape its narrative. The detection of its moon, Charon, in 1978 revealed a companion so massive that the two bodies orbit a shared center of gravity outside Pluto itself, marking them as a unique binary system. Later observations identified smaller moons—Styx, Nix, Kerberos, and Hydra—adding to the complexity of Pluto's environment.

The realization that Pluto was not alone in its orbit but part of a vast region of icy bodies transformed its status. The discovery of the Kuiper Belt in the 1990s reframed Pluto not as an anomaly but as a representative of a new class of objects. This recognition culminated in 2006 with the International Astronomical Union's controversial decision to reclassify Pluto as a dwarf planet. While this change sparked heated debates and public outcry, it reflected the evolving nature of science—a discipline that embraces complexity and adjusts its models to accommodate new evidence.

Pluto's reclassification also echoes its mythological associations with liminality and ambiguity. It defies easy categorization, existing simultaneously as a dwarf planet, a Kuiper Belt object, and a former member of the solar system's planetary family. Its demotion challenges humanity's desire for neat definitions and clear boundaries, reminding us that the universe is inherently diverse and resists simplification.

The exploration of Pluto by the New Horizons spacecraft in 2015 was a turning point in its scientific story. For decades, Pluto had been a distant dot, its surface and nature shrouded in mystery. The arrival of New Horizons changed everything. High-resolution images revealed a world far more dynamic and complex than anyone had imagined. The vast, heart-shaped plain of Tombaugh Regio, the towering water-ice mountains, and the evidence of active geological processes all pointed to

Pluto

a planet that was very much alive. Pluto was not a frozen relic but a dynamic world shaped by internal and external forces, its surface continually reshaped by the interplay of nitrogen ice, methane frost, and atmospheric escape.

Tombaugh Regio, with its smooth plains and convection cells, became an emblem of Pluto's story. Its heart shape, visible even in early images, captured the public's imagination, transforming Pluto from a scientific curiosity into a world with character and charm. This feature, named after its discoverer, symbolizes the emotional and intellectual resonance of Pluto's discovery and exploration.

Pluto's story is one of transformation and complexity, a narrative that reflects the human capacity for discovery and redefinition. Its dual identity as both a mythological and scientific entity underscores the interplay between imagination and reason in our quest to understand the cosmos. The name Pluto, chosen to honor a figure of mystery and transition, reminds us that the universe is both a source of inspiration and a realm of empirical exploration.

As we study Pluto, we are reminded of the limits of categorization and the importance of embracing diversity. Pluto challenges us to think beyond rigid definitions, to see the solar system not as a collection of isolated objects but as a dynamic and interconnected system. It invites us to celebrate the richness of the cosmos, where every object, no matter how distant or small, contributes to the greater narrative of existence.

Pluto's enduring appeal lies in its ability to bridge the ancient and the modern, the mythical and the empirical. It is a world that invites reflection on the nature of exploration, identity, and the boundaries of knowledge. In Pluto, we find a celestial body that is as much a mirror of ourselves as it is a window into the universe—a reminder that even the farthest reaches of the solar system are filled with wonder and meaning.

Chapter 7: The Science of the Kuiper Belt

The Kuiper Belt is one of the most fascinating and enigmatic regions in our solar system. It lies beyond Neptune, stretching from about 30 AU (astronomical units) to roughly 50 AU from the Sun, and serves as a cosmic frontier, a distant realm that remains largely unexplored and full of secrets. The discovery of the Kuiper Belt reshaped our understanding of the solar system and its formation, as well as our place within the larger context of the cosmos.

In the decades following Pluto's discovery in 1930, astronomers debated whether there were other objects beyond Neptune that could have similar properties to Pluto. The idea was mostly speculative until the 1990s, when the existence of the Kuiper Belt was confirmed. The region, named after the Dutch-American astronomer Gerard Kuiper, who had proposed its existence in the 1950s, is populated by a wide variety of icy bodies, many of which are remnants from the early solar system. These objects, often referred to as Kuiper Belt Objects (KBOs), range in size from small icy rocks to large dwarf planets like Pluto itself. The Kuiper Belt is sometimes described as the solar system's "third zone," following the inner planets and the gas giants, but it is far more than just a void or a distant expanse. It is a dynamic, evolving area that holds clues to the very origins of the solar system.

The Kuiper Belt objects are composed primarily of ices—water, ammonia, and methane—mixed with rock. These icy bodies are often referred to as "frozen relics," because they are considered to be leftover materials from the solar system's formation. Unlike the planets, which were formed from a protoplanetary disk of gas and dust, the objects in the Kuiper Belt likely formed from a mixture of ices and rocks that never coalesced into planets. In many ways, they are like fossils, preserving the conditions of the solar system during its formative years. The Kuiper Belt is thus a time capsule, holding pieces of the ancient solar nebula that existed before the planets fully formed.

Pluto

The significance of the Kuiper Belt goes far beyond its role in the formation of the solar system. It provides us with a direct link to the outermost reaches of our planetary neighborhood, where the Sun's influence wanes and the cold of interstellar space begins to assert itself. The region is believed to be the source of many short-period comets—those that have orbits of less than 200 years. These comets, when they swing into the inner solar system, bring with them primordial material that can help us learn about the composition and evolution of the early solar system. Studying these comets is crucial for understanding the conditions that led to the formation of Earth and the other planets.

The first confirmed Kuiper Belt Object, 1992 QB1, was discovered by astronomers David Jewitt and Jane Luu in 1992, a finding that ignited a new era of exploration. Before this discovery, Pluto had been considered an anomaly—an isolated object far out in the solar system, possibly the last of its kind. But 1992 QB1 and other similar objects demonstrated that Pluto was not alone. There was an entire population of similar objects, ranging in size from a few kilometers to several hundred kilometers in diameter. These objects are located in a region that is not densely populated but is still large enough to contain a vast array of mysterious objects, many of which are still being discovered today.

As our understanding of the Kuiper Belt grew, so too did our understanding of the complexity and diversity of its inhabitants. Some of these objects, like Pluto, are large enough to be classified as dwarf planets. Others are much smaller and remain little more than frozen clumps of ice and rock. But what makes the Kuiper Belt so intriguing is its variety. It's not just a collection of icy bodies; it's a dynamic region with objects that exhibit a wide range of behaviors. Some are in highly elliptical orbits that take them close to the Sun and then sling them back out to the farthest reaches of the solar system, while others are locked in more stable, circular orbits. The Kuiper Belt is also home to some of the solar system's most peculiar objects, such as the scattered disk—an area just beyond the Kuiper Belt where objects have been flung into highly eccentric orbits by the gravitational forces of Neptune.

One of the most surprising aspects of the Kuiper Belt is the realization that it may not just be a relic of the solar system's past but also a key player in the solar system's future. Objects from the Kuiper Belt have been ejected outward by the gravitational forces of the giant planets and could potentially enter interstellar space. Meanwhile, other objects from the Kuiper Belt have been gravitationally captured by the giant planets and could potentially become moons of the gas giants. This interplay between the Kuiper Belt and the rest of the solar system adds a layer of dynamism and complexity to the region, transforming it from a static relic into an active, evolving zone.

In addition to Pluto, other large bodies in the Kuiper Belt, such as Eris, Haumea, and Makemake, have been classified as dwarf planets. These objects have provided further insights into the structure and nature of the Kuiper Belt. Their irregular shapes, unique compositions, and distinctive features have forced scientists to reconsider the traditional boundaries between planets, dwarf planets, and small solar system bodies. The discovery of these objects, combined with the study of comets and other Kuiper Belt objects, has led to the realization that the solar system is far more dynamic and complex than previously thought.

The study of the Kuiper Belt has also led to a greater understanding of the formation of planetary systems beyond our own. Many of the objects in the Kuiper Belt are similar in composition to the objects that are thought to make up the disks surrounding other stars. The study of these objects has therefore provided valuable insights into the processes that shape planetary systems across the galaxy. Moreover, the discovery of exoplanets—planets that orbit stars other than the Sun—has also contributed to our understanding of the Kuiper Belt's role in the broader context of planetary formation. As we explore the Kuiper Belt, we are learning not only about our own solar system but also about the broader processes that shape the evolution of planets and stars throughout the universe.

In the modern era of space exploration, the Kuiper Belt has become a major focus of scientific study. NASA's New Horizons mission, which made its historic flyby of Pluto in 2015, has expanded our knowledge of the Kuiper Belt and the outer solar system. New Horizons has revealed unprecedented details about Pluto's surface, atmosphere, and moons, providing scientists with a wealth of information about this distant dwarf planet. The spacecraft is continuing its journey outward, and in

Pluto

2019, it made a flyby of another Kuiper Belt object, Arrokoth, providing the first close-up images of a Kuiper Belt object.

The Kuiper Belt remains an area of great mystery and immense potential. As we continue to explore this distant region, we are bound to uncover more secrets about the early solar system, the nature of icy bodies, and the evolution of planetary systems. Just as Pluto continues to captivate our imagination, the Kuiper Belt, with its hidden treasures and unexplored reaches, promises to reveal much more about the origins and future of the solar system.

In the grand scheme of the solar system, the Kuiper Belt represents both an end and a beginning. It is the last frontier, the final domain of our solar system before the unknown reaches of interstellar space. But it is also a doorway, a gateway to greater understanding of the cosmos. As we explore the Kuiper Belt, we are not simply charting a distant region; we are tracing the roots of our own existence and expanding our view of the universe itself.

Chapter 8: The Limits of Categorization

The story of Pluto's demotion from a planet to a "dwarf planet" encapsulates much more than just the evolution of an astronomical classification; it touches on the very nature of how humanity approaches knowledge, categorization, and understanding. The decision by the International Astronomical Union (IAU) in 2006 to officially strip Pluto of its planetary status was both an act of scientific precision and a cultural earthquake, forcing us to question the rules we use to define and classify the natural world.

For centuries, the notion of what constitutes a planet was simple and largely unquestioned. For most of human history, planets were simply those objects that wandered across the sky. The word "planet" itself comes from the ancient Greek word *planētēs*, meaning "wanderer," a reference to their movement among the fixed stars. In the early days of astronomy, a planet was something you could see moving, and this was enough to define it. As our understanding of the solar system grew, we refined our definition. Planets were those objects that orbited the Sun, were spherical in shape due to their self-gravity, and had cleared their orbits of other debris. This definition served for centuries, providing a stable framework in which our understanding of the cosmos was built.

But science is not static, and our definitions of concepts like "planet" are subject to revision. The discovery of Pluto in 1930 was an important moment in our understanding of the solar system. At the time, Pluto seemed to fit perfectly into the established definition of a planet. It orbited the Sun and had the spherical shape required by the existing model. Yet, as telescopes became more powerful, astronomers began to realize that Pluto's orbit was unlike those of the other planets. Its path was highly elliptical and tilted relative to the plane of the solar system. And, as more objects were discovered in the same region beyond Neptune, it became apparent that Pluto might not be alone in its orbit. There was a growing realization that Pluto might simply be one of many small icy bodies in a newly discovered region: the Kuiper Belt.

Pluto

By the time the International Astronomical Union convened in 2006 to discuss the status of Pluto, astronomers had discovered hundreds of similar objects in the Kuiper Belt, including Eris, a dwarf planet that was slightly more massive than Pluto. The question arose: if Pluto was a planet, should these other objects also be considered planets? To resolve this, the IAU established a new definition of a planet, one that required an object not only to orbit the Sun and be spherical in shape, but also to have "cleared the neighborhood" around its orbit—meaning it must be dominant in its path, without sharing it with other comparable-sized bodies. Pluto, with its highly eccentric orbit and location in the crowded Kuiper Belt, did not meet this criterion.

The demotion of Pluto was met with an outcry from many who felt that this decision was arbitrary, unnecessary, and unfair. To some, Pluto was still a planet, regardless of the new definition. After all, Pluto had held that title for over 75 years, and the emotional attachment many felt to it as the ninth planet in the solar system was hard to shake off. Moreover, there was the sense that this reclassification was an oversimplification, that it stripped Pluto of its uniqueness and specialness by reducing it to just another icy object in the Kuiper Belt.

This debate raises larger philosophical questions about categorization itself. The act of classifying things into neat, tidy categories is one of humanity's greatest intellectual achievements. It allows us to make sense of the world, to impose order on complexity. But categorization can also be limiting, restricting, and even arbitrary. Nature, as it turns out, often resists such neat divisions. In many ways, the debate over Pluto's status is a reflection of the limitations of our language and frameworks for understanding the universe. Categories like "planet," "dwarf planet," and "moon" are simply human constructs, and they only partially capture the full diversity of the objects they are meant to describe.

For instance, consider the case of Eris. Though slightly more massive than Pluto, it is still a member of the same category of objects that are mostly composed of ice and rock, residing in the outer regions of the solar system. Eris shares much in common with Pluto, yet it is labeled as a "dwarf planet" rather than a full-fledged planet. The line between planet and dwarf planet is thin, perhaps thinner than we would like to admit. And what about Ceres, the dwarf planet in the asteroid belt? For centuries, Ceres was classified as a planet, before being reclassified as an asteroid. It was only in 2006, when it was recognized as a spherical object in its own right, that Ceres was again classified as a dwarf planet. But this raises the question: why is the distinction between a planet and a dwarf planet so important? And what, ultimately, does it mean for something to be a "planet"?

The struggle with categorization is not limited to just planetary science. It reflects a broader tension in science and philosophy between the desire for clarity and the acknowledgment of complexity. We long for certainty, but the universe itself is often elusive and resistant to our attempts to impose rigid boundaries. The case of Pluto highlights the ways in which our definitions evolve and adapt as we learn more about the world around us. Pluto's story is a metaphor for the ongoing process of scientific discovery: as we venture deeper into the unknown, we find that our initial frameworks may no longer be sufficient to explain what we encounter. The universe is far more complicated than we imagined, and our classifications must evolve with it.

In many ways, Pluto's demotion can be seen as an example of how science continuously evolves and refines its understanding of the natural world. It is an essential aspect of scientific inquiry that classifications, definitions, and even the very boundaries of knowledge are subject to change. The decision to reclassify Pluto was based on new discoveries, new methods of observation, and a deeper understanding of the solar system. But this evolution also points to the limitations of human knowledge. We may never fully grasp the extent of the universe, but as we press forward into new territories, we can take solace in the fact that our definitions, though imperfect, are an attempt to grasp something far greater.

The categorization of Pluto, then, serves as a reminder that science is not just about uncovering facts but about refining our models, adjusting our perspectives, and acknowledging the inherent complexity of the universe. It challenges us to think critically about how we define and understand the world around us. Pluto may no longer be a planet in the strictest sense, but its role in our imagination, our exploration, and our understanding of the solar system is far from diminished. Whether classified as a planet, a dwarf planet, or simply an object in the outer reaches of the solar system, Pluto continues to captivate our curiosity and inspire wonder. Its story, like the stories of all the planets and moons, is part of the larger narrative of exploration—a narrative that is constantly evolving, just as our understanding of the universe does.

Ultimately, Pluto's demotion is less about the object itself and more about the process of redefinition—how science constantly tests the limits of categorization and challenges our perceptions of reality. Pluto remains, in many ways, an icon of our journey into the unknown, a symbol of humanity's relentless desire to understand the cosmos, no matter how difficult or elusive the answers may be.

Conclusion: The Heart of the Outer Reaches

Pluto stands at the threshold of the unknown, a boundary marker between the known universe and the vast, mysterious realms that lie beyond. Its reclassification, though controversial, serves as a profound reminder of how our understanding of the cosmos continues to evolve. In its quiet exile on the fringes of the solar system, Pluto has come to symbolize much more than a mere astronomical object. It has become an emblem of humanity's ongoing struggle to define and comprehend the cosmos, an embodiment of both the potential and the limitations of our scientific endeavors.

As we reflect on Pluto's journey—from its discovery in 1930, to its exaltation as the ninth planet, to its subsequent demotion—there is an underlying message: the universe is far more complex and nuanced than we could have ever imagined. Our definitions, though vital for organizing knowledge, are ultimately human constructs, subject to revision as new information becomes available. Pluto's story is not just one of a planet stripped of its title but of the larger, ever-changing narrative of science itself—a narrative that embraces uncertainty, challenges assumptions, and adapts in the face of discovery.

The exiled wanderer of the Kuiper Belt teaches us a profound lesson about our own place in the universe. It serves as a reminder that our understanding of the solar system—and the cosmos at large—is still in its infancy. We are explorers, not conquerors of knowledge, and like Pluto, we too are on a journey that will likely never have a final, definitive destination. As we venture further into the outer reaches of space, our definitions and categories will inevitably continue to shift. We will discover new objects, new phenomena, and perhaps even new laws of physics that will force us to revise what we know. In this sense, Pluto's demotion is not a conclusion, but a prelude—an indication that our journey into the unknown has only just begun.

This distant world, far from being relegated to obscurity, has captured the imagination of generations. Its small, icy heart has sparked debates

Pluto

about the nature of exploration and discovery, about the human impulse to classify, and about the need for definition in a universe that often defies easy categorization. Pluto has become, in many ways, a symbol of our yearning to explore what lies at the edges of our understanding. Despite being on the fringes of the solar system, Pluto has played a central role in expanding our vision of the cosmos. Its existence has propelled us to look beyond the familiar planets, to peer into the mysterious regions of the Kuiper Belt, and to reconsider how we define a planet.

In the same way that Pluto exists in the space between the known and the unknown, we as humans exist in the space between certainty and uncertainty. Our quest for knowledge is driven by an unrelenting desire to push the boundaries of what we know. This quest is never about reaching a final answer but about constantly asking questions, redefining categories, and exploring deeper and deeper into the vastness of space.

It is fitting, then, that Pluto occupies the furthest reach of the solar system. It is a planet that has long stood on the edge of human knowledge, a world that beckons us to venture further, to look closer, and to think more deeply about the universe and our place within it. Its icy, enigmatic heart holds the secrets of the outer solar system—secrets that we may not fully uncover for generations to come. And yet, in the search for these secrets, we are bound to encounter new insights not just about the distant worlds of our solar system, but about the very nature of science itself.

The lessons Pluto imparts are not limited to its physical properties or the mysteries of its surface. More profoundly, Pluto teaches us about the nature of human knowledge: that it is always provisional, always incomplete, and always subject to change. Pluto's journey mirrors our own: as we explore the unknown, we are constantly revising our definitions, constantly rethinking our assumptions, and constantly finding that the universe is far more complicated, more surprising, and more beautiful than we could ever have imagined.

What does Pluto's story say about us? It speaks to our drive to explore, to our need to categorize, and to our deep-seated curiosity about the cosmos. But it also speaks to our limitations—to our inability to fully understand the vast complexity of the universe and our tendency to make sense of it through the tools at our disposal. Pluto's reclassification, though contentious, is a testament to the growth of human understanding. We may not yet have all the answers, but each new discovery, each new classification, brings us closer to the heart of the cosmos.

Ultimately, Pluto remains the exiled wanderer, a symbol of humanity's reach into the unknown and of the mysteries that still await. It is a planet, and a dwarf planet, and yet it is also something more: a reminder of the limits of our knowledge, the fluidity of our definitions, and the endless frontier that lies beyond. As we look outward, beyond the planets we know and into the distant reaches of the solar system and beyond, Pluto reminds us that the journey is as important as the destination, and that in the search for answers, the questions are just as valuable as the solutions.

In the end, Pluto will always hold a special place in our hearts—not because of the controversy over its classification, but because it represents the spirit of discovery. It is a world that has beckoned us to look farther, to think deeper, and to question what we think we know. It may be the farthest planet from the Sun, but it is also the closest to our dreams, to our endless curiosity, and to our unyielding desire to understand the cosmos.

End Note: The Wanderer's Tale

Pluto's story is not just one of scientific discovery, but one of transformation, both for the celestial body itself and for humanity's understanding of its place in the universe. Its journey from discovery to planetary status, then to its controversial reclassification as a "dwarf planet," serves as a microcosm of our evolving relationship with the cosmos. It is a reminder that our concepts of space, classification, and knowledge are not fixed but ever-changing, driven by new discoveries, evolving technologies, and an unyielding curiosity to explore what lies beyond our understanding.

In the grand scheme of our solar system, Pluto is but one small object—distant, icy, and enigmatic. Yet, as we've learned from the New Horizons mission and subsequent scientific research, it is also a place of surprising complexity, offering insights into the processes that govern the outer reaches of our solar system. From its mountainous terrains and vast nitrogen ice plains to its surprising geological activity, Pluto continues to defy expectations, providing us with new questions and mysteries to unravel. It is no longer just a frozen rock at the edge of the solar system, but a world in its own right, brimming with secrets waiting to be discovered.

Yet, Pluto is not alone in its position at the fringes of our solar system. It resides in the Kuiper Belt, a vast and largely unexplored region that is home to countless icy bodies, many of which have yet to be fully understood. This region, which stretches far beyond the orbit of Neptune, has become a new frontier in planetary science, and Pluto, though no longer a "planet," continues to play a pivotal role in our exploration of this distant zone. Its status as a dwarf planet has only increased its mystique, positioning it as a symbol of the uncharted and the unknown. And in this role, Pluto stands as a beacon, guiding us toward further exploration of the distant reaches of our solar system and beyond.

In contemplating Pluto's significance, it is helpful to step back and reflect on the broader narrative of exploration that stretches across our entire solar system. Pluto's story is inextricably linked to the stories of the nine other planets we've encountered so far—each offering unique lessons

Pluto

about the diversity, beauty, and complexity of the cosmos. Our exploration of these worlds—from the searing surface of Mercury to the stormy skies of Neptune—has taught us not only about the planets themselves but also about the very nature of science, discovery, and human curiosity.

www.ingramcontent.com/pod-product-compliance
Lightning Source LLC
Chambersburg PA
CBHW070944220526
45469CB00007B/2506